Learn About

FALL

PUMPKINS

BY BRENNA MALONEY

Children's Press®
An imprint of Scholastic Inc.

Library of Congress Cataloging-in-Publication Data available

ISBN 978-1-5461-0178-9 (library binding) | ISBN 978-1-5461-0180-2 (paperback)

10 9 8 7 6 5 4 3 2 1   25 26 27 28 29

Printed in China 62
First edition, 2025

Book design by Kay Petronio

Photos ©: 5 top left: Lolly12gate/Dreamstime; 5 top right: dial-a-view/Getty Images; 6 top: Spauln/Getty Images; 6 bottom: EvgeniiAnd/Getty Images; 7 bottom left: Chiyacat/Getty Images; 8: igor_kell/Getty Images; 9 top: sanapadh/Getty Images; 12-13: redmal/Getty Images; 14 bottom: anmbph/Getty Images; 15: precinbe/Getty Images; 18 bottom: by John Carleton/Getty Images; 21: Kingfisher Productions/Getty Images; 22 bottom: KirbyIng/Getty Images; 23 top: Dar1930/Getty Images; 23 bottom: etorres69/Getty Images; 28 raw seeds: sanapadh/Getty Images; 29 center right: Kelly Sillaste/Getty Images; 30 top right: Terry Schmitt/UPI/Shutterstock; 30 right center top: Jason Alig/Sidney Daily News/AP Images; 30 right center bottom: Wang Kai/Xinhua/Getty Images; 30 bottom: Seth Perlman/AP Images.

All other photos © Shutterstock.

CONTENTS

Introduction

WELCOME TO THE PUMPKIN PATCH!

Look at all the pumpkins!

Fall is **harvest** time for pumpkins. This means they are ready to be picked!

Have you ever picked your own pumpkin?

4

Sugar
pumpkin

Fairytale
pumpkin

Full Moon
pumpkin

Mellow Yellow
pumpkin

Batwing
pumpkin

On the outside, pumpkins can be different colors.

Most are orange. Some are white, yellow,
or green.

Some pumpkins are small. They fit in your hand.

Others are big. You need a wagon to carry them.

A large pumpkin can weigh about 30 pounds (14 kg). That is around the same weight as two bowling balls!

Pumpkins can be thin or round.

They can be smooth or bumpy.

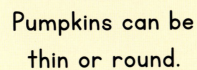

There are about 150 different types of pumpkins.

Look inside! A pumpkin has pulp.

The center is filled with slimy, stringy strands.

Stuck to all the strings are seeds.

You can cook and eat the pulp of a pumpkin.

pulp

Pumpkin seeds can be roasted and eaten, too!

seed

A pumpkin seed is flat. It is shaped like a teardrop.

A pumpkin might have as many as 500 seeds inside!

Chapter 1
PLANTING

A pumpkin begins as a seed.

It is spring. A seed is planted in the dirt.

The best time to plant pumpkin seeds is in late May or early June.

Inside, a pumpkin seed is green. The seed is covered by a white shell.

Pumpkins take time to grow.

They need water and sunlight.

Watch and wait. After about 10 days, *POP!*

A tiny green plant breaks through the dirt. This is called the shoot.

POP

POP

shoot

seed

roots

Below the ground, the pumpkin plant has **roots**. They are growing, too.

Seven days later, two small leaves open.

stem

They are held up by a skinny **stem**.

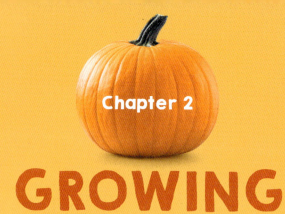

Chapter 2

GROWING

During summer, the pumpkin plant will grow and grow.

It will become a **vine**. The vine will creep along the ground.

↑
vine

Pumpkin vines can grow to be 20 to 30 feet (6 to 9 m) long. That is about the same length as a garden hose.

Pumpkins are fruits! That is because they grow from the flowers of a plant.

Big green leaves will grow on the vine.

Wait about eight weeks. Bright yellow flowers start to bloom.

In the place of each flower, a pumpkin will begin to grow.

At first, it is very small.

The small, green ball in this picture is the start of a pumpkin!

Pumpkin leaves usually have five to seven points.

A pumpkin is full-grown around 120 days after a seed is planted.

By the end of the summer, the pumpkin is much bigger!

It is about the size of a basketball.

Chapter 3

PICKING AND EATING

At last, it is fall! Harvest time! Is the pumpkin ready?

When an orange pumpkin is ripe, it has a bright color. Its stem is brown.

stem

Pumpkin stems are also called handles.

Tap the pumpkin. What sound does it make?

Thump, thump. It is ready to be picked!

More than 800 million ripe pumpkins are picked in October in the United States.

These pumpkins have not been picked yet. They are still attached to their vines.

Farm stands usually have different types of pumpkins to choose from.

Where can you find pumpkins to buy?
At a farm stand. In a grocery store.
At a pumpkin patch.

$2.00 or 3 for $5.00

It is hard to just pick one!

Bring your pumpkins home. What's next?

About 1.5 billion pounds of pumpkins are sold in the United States every year!

Pumpkins are a healthy food. They are full of **vitamins**.

Pumpkins are also tasty!

Bake a pumpkin pie. *Yum!* Use spices like cinnamon and nutmeg.

cinnamon

nutmeg

You can make a pumpkin pie with fresh or canned pumpkin.

Warm pumpkin soup can be served inside a pumpkin shell.

Or cook some pumpkin soup. Or make pumpkin bread.

This loaf of pumpkin bread is topped with pumpkin seeds.

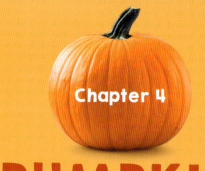

Chapter 4
PUMPKIN FUN

You can change the way your pumpkin looks!

Paint a face on the outside.

It can be silly. Or scary! *Boo!*

Little lightbulbs, called tea lights, with batteries are a safe way to light your jack-o'-lantern.

A carved pumpkin will start to soften in three to five days.

Carve your pumpkin into a jack-o'-lantern.

Put a little light inside. Watch it glow!

Why are pumpkins so popular in fall?

Pumpkins are ripe in the fall. They are picked and ready to be eaten!

We decorate with jack-o'-lanterns at Halloween.

Jack-o'-lanterns started in Ireland. Instead of pumpkins, people carved turnips and potatoes!

The most popular jack-o'-lanterns have faces!

Some people decorate tables with pumpkins and corn in the fall.

Pumpkin pie is eaten at Thanksgiving.

What will you do with your pumpkin this fall?

MAKE ROASTED PUMPKIN SEEDS

You can roast pumpkin seeds for a tasty snack!

YOU WILL NEED:

An adult helper

2 cups of raw pumpkin seeds

A cookie sheet

Paper towels

1 tablespoon olive oil

Water for cleaning

1 tablespoon salt

STEPS:

1. Have an adult preheat the oven to 350°F (177°C).

2. Wash the seeds to remove any leftover pumpkin.

3. Dry the seeds on paper towels.

4. Spread the seeds on the cookie sheet.

5. Coat the seeds with the olive oil and sprinkle with the salt. Mix with clean hands.

6. Bake for 30 minutes, or until they reach a light golden-brown color.

7. After your seeds cool off, enjoy!

PUMPKIN FACTS

The world's heaviest pumpkin weighed 2,749 pounds (1,247 kg).

Travis Gienger grew the world's heaviest pumpkin in Minnesota.

The world's largest pumpkin pie weighed 3,699 pounds (1,678 kg).

It took 1,212 cans of pumpkin to make this pie!

China produces the most pumpkins in the world. More than 18 billion pounds (8 billion kg) of pumpkins are grown each year!

More pumpkin is eaten in China than in any other country in the world.

The town of Morton, Illinois, calls itself the Pumpkin Capital of the World. Morton produces 85 percent of the world's canned pumpkin.

A field of pumpkins in Morton, Illinois.

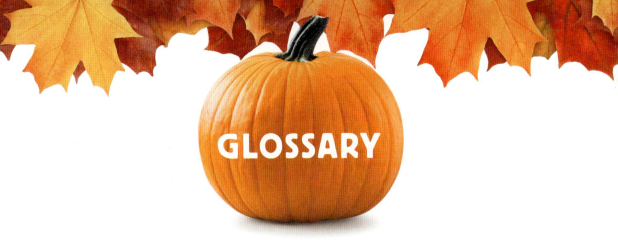

GLOSSARY

harvest (HAHR-vist) to gather crops from a field

pulp (puhlp) the soft, juicy part of fruits and vegetables

root the part of a plant or tree that grows under the ground, where it collects water and nutrients

seed the part of a flowering plant from which a new plant can grow

stem the main, upward-growing part of a plant from which the leaves and flowers grow

vine a plant with a long stem that grows along the ground

vitamin (VYE-tuh-min) one of the substances in food that is essential for good health and nutrition

INDEX

ABOUT THE AUTHOR

BRENNA MALONEY is the author of many books. She lives in Washington, DC, with her husband and two sons. She loves carving jack-o'-lanterns and eating roasted pumpkin seeds.